Cloud cover

Cloud cover

Measuring the Weather

Alan Rodgers and Angella Streluk

Heinemann Library
Chicago, Illinois

Published by Heinemann Library,
an imprint of Reed Educational & Professional Publishing,
Chicago, Illinois
Customer Service 888-454-2279
Visit our website at www.heinemannlibrary.com

Design by Storeybooks
Originated by Ambassador Litho Limited
Printed in Hong Kong/China

07 06 05 04 03
10 9 8 7 6 5 4 3 2

Library of Congress Cataloging-in-Publication Data

Rodgers, Alan, 1958-
Cloud cover / Alan Rodgers and Angella Streluk.
v. cm. -- (Measuring the weather)
Includes bibliographical references and index.
Contents: No sun, no weather! -- What is the sun? -- The sun and temperature -- Day and night -- UV radiation -- What are clouds? -- Cloud cover -- Cloud classification -- Cumulus clouds -- Stratus clouds -- Cirrus clouds -- Giant clouds -- Fog and visibility.
ISBN 1-58810-686-1 -- ISBN 1-40340-126-8 (pbk.)
1. Cloudiness--Juvenile literature. 2. Clouds--Juvenile literature.
[1. Clouds.] I. Streluk, Angella, 1961- II. Title.
QC921.35 .R64 2002
551.57'6--dc21

2002004021

Acknowledgments

The author and publishers are grateful to the following for permission to reproduce copyright material: The Art Archive, p. 4; Robert Harding Picture Library, pp. 6, 25; Stone, p. 7; Trevor Clifford Photography, pp. 8, 15, 16; Eye Ubiquitous/NASA, p. 10; Bruce Coleman Collection, pp. 11, 24; Topham Picturepoint, p. 13; K. E. Woodley/The Met Office, p. 18; GeoScience Features, pp. 20, 22, 26 (pictures 1, 2, 4); Science Photo Library, pp. 21, 26 (pictures 3, 6); R. N. Hughes/The Met Office, p. 23; T. J. Lawson/The Met Office, p. 26 (picture 5); FLPA, p. 28.

Cover photographs reproduced with permission of Science Photo Library and Photodisc.

Our thanks to Jacquie Syvret of the Met Office for her assistance during the preparation of this book.

Some words are shown in bold, **like this.** You can find out what they mean by looking in the glossary.

Contents

No Sun, No Weather!

It is very easy to take the Sun for granted, but to do so is a mistake. The Sun is the power behind the weather. Without it there would be no sunshine, rain, wind, fog, or hail. In fact, life on Earth would end.

The Sun is also important because it causes day and night. It also causes the seasons. This means that each area of the world has weather patterns that repeat themselves again and again. This gives each place its different **climate.** Weather is measured from day to day, whereas climate is the pattern of the weather over a longer period of time.

It is important to know about the Sun because it affects the way we live. Also, as we shall see, it plays a part in the formation of clouds.

Artists have often shown clouds in their paintings. The artist John Constable was famous for his paintings of skies. In this painting—called *Chair Pier, Brighton*—the sky was given as much space as the rest of the picture put together.

Clouds

To many people, clouds are a mystery. They think that learning to identify them would be too complicated. This is not true. There are simple ways of sorting clouds to help identify them. Once you know what the different clouds are, you will be able to find out what kind of weather they are likely to bring.

Weather symbols

In professional weather reports, the weather is shown on maps with a set of symbols that are internationally recognized. This means that **data** can be shared by **meteorologists** around the world. By sharing data, a larger picture of the weather can be seen. This means that we can predict it more efficiently. Sometimes different symbols are used for weather reports on television, in newspapers, and on the Internet. These are more like pictures and can easily be understood by the public.

Be careful!

Do not look directly at the Sun when studying the weather, and never seek shelter under trees during a thunderstorm, as they may be hit by lightning.

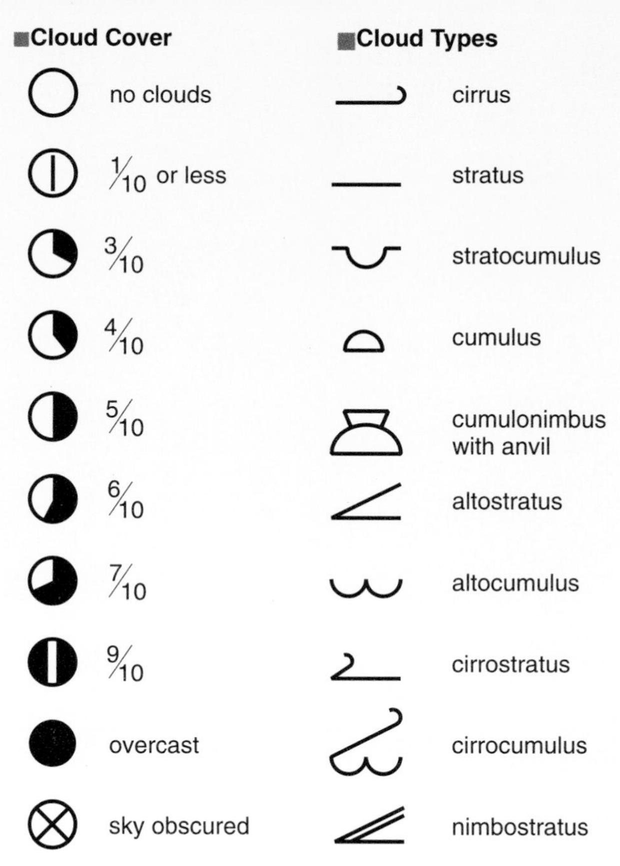

The drawings on the left show how much of the sky is covered by clouds. Some people measure this in eighths. Other people measure this in tenths, as seen here.

The drawings on the right represent the different kinds of clouds.

What Is the Sun?

Where does the Sun come from? Scientists think our universe was made through a process called the **big bang,** between ten and twenty billion years ago. When this happened, the Sun was formed out of billions of swirling particles of dust. The Sun is approximately five billion years old and has enough energy to last another five billion years. Fortunately, this means that the Sun will be with us for a long time.

The Sun is the source of all our natural light. Without it, we would all freeze to death. Sunlight is also reflected off the moon at night, giving moonlight. The Sun looks relatively big to us because it is only 93 million miles (150 million kilometers) away. It only takes eight minutes for light from the Sun to reach us. Compared to this, light from the nearest bright star—Alpha Centauri A—takes four years to reach us, even though light travels at 186,282 miles (299,972 kilometers) per second!

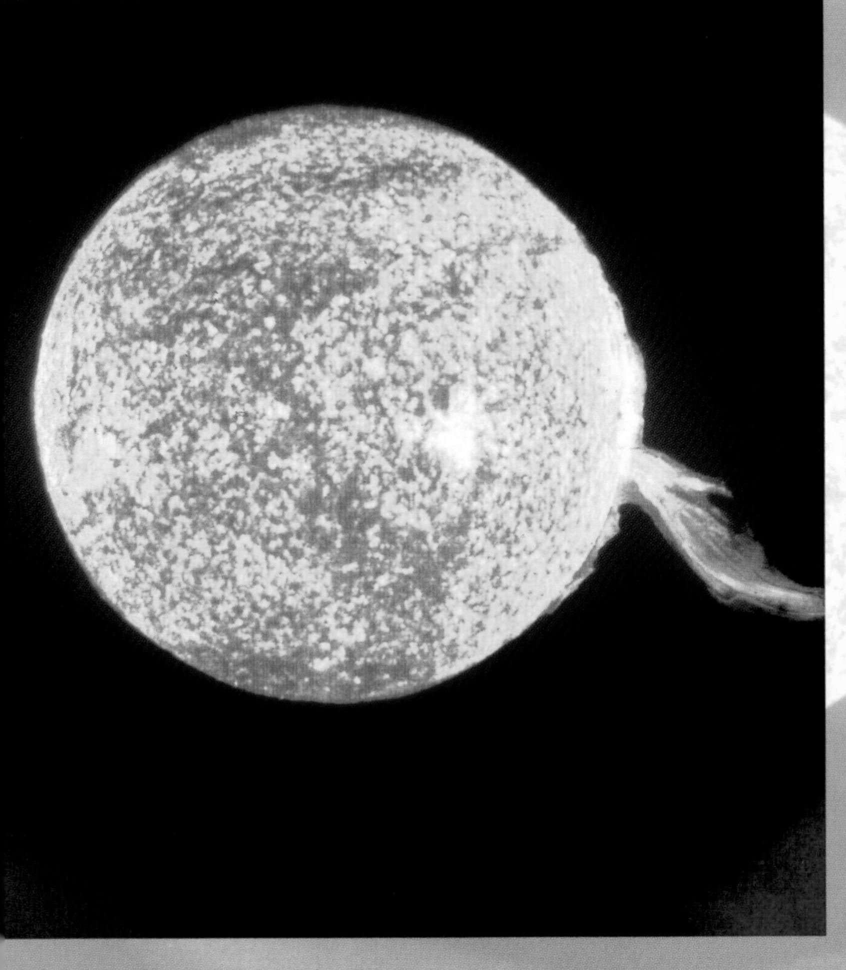

Did you know that our Sun is about 864,000 miles (1.4 million kilometers) in diameter? It is about five billion years old, and the temperature at its core is over 27 million °F (15 million °C).

The Sun is like a giant engine that causes our weather. As the earth rotates, the half that faces the Sun warms up. The other half cools down. Because the earth is curved—like a ball—some parts, especially the **equator,** are warmed up more than others because they stick out farther. Land and sea also absorb heat at different speeds. All of these things cause hot air to rise and cold air to sink in different places and at different times. This process gives us our winds.

Thermals

When the Sun heats up the earth, it can cause heat to rise in currents called **thermals.** On warm, sunny days, these produce **cumulus** clouds (see pages 20–21). As the Sun heats the air, moisture rises in the sky until it cools and condenses. This produces clouds that look very white and bright in the sunshine because the moisture reflects sunlight very well. The fluffy cumulus clouds do not last for a very long time.

Like some birds, this glider uses air currents called thermals to give it lift. It does not have an engine to help lift it higher into the sky. The pilot looks for telltale signs of thermals, such as fluffy cumulus clouds. Pilots also know where thermals are most likely to rise. These thermals help keep gliders in the air.

The Sun and Temperature

The temperature is not the same all around you. Even within a small area, temperatures are different. The largest differences are between areas of shade and direct sunshine. This is why professional **meteorologists** always measure temperatures in the shade. They can then compare their **data** knowing that they are all measuring it in the same way. Clouds form their own shade, which reduces the temperature.

The Sun causes changes in temperature. About 19 percent of the heat from the Sun is absorbed by the **atmosphere** and clouds. Some 51 percent of the heat is absorbed into the earth's surface. About 30 percent of all the heat from the Sun is reflected back into space.

The earth is a **sphere,** which means that the Sun's rays fall more directly onto the **equator.** Towards the **poles,** the curve of the earth means the rays strike it at an angle. The rays become more spread out, and the heat is less intense. Where the Sun's rays are more concentrated—like at the equator—it will be hotter.

You can use a flashlight and a globe to show the different amounts of sunlight that different parts of the world receive. When the flashlight is shone onto the nearest part of the globe, the circle of light will be small. If the light is shone near the top of the globe, it will become a shape called an **ellipse.** The light leaving the flashlight is the same, but the area it is spread over gets bigger or smaller. This is also the way that sunlight shines on the earth.

The greenhouse effect

Even though much heat is lost through the atmosphere, the earth stays warm because of the **greenhouse effect.** Special gases called greenhouse gases keep heat in the atmosphere. Without these gases, the temperature on earth would be much colder than it is. Many scientists are worried that humans have increased the amount of greenhouse gases in the atmosphere. This means more heat has been trapped, upsetting the delicate balance of earth's atmosphere.

Carbon dioxide is one of the greenhouse gases. It is released when **fossil fuels** are burned. The gases used in some refrigerators and aerosol cans have also been linked to the increase in greenhouse gases.

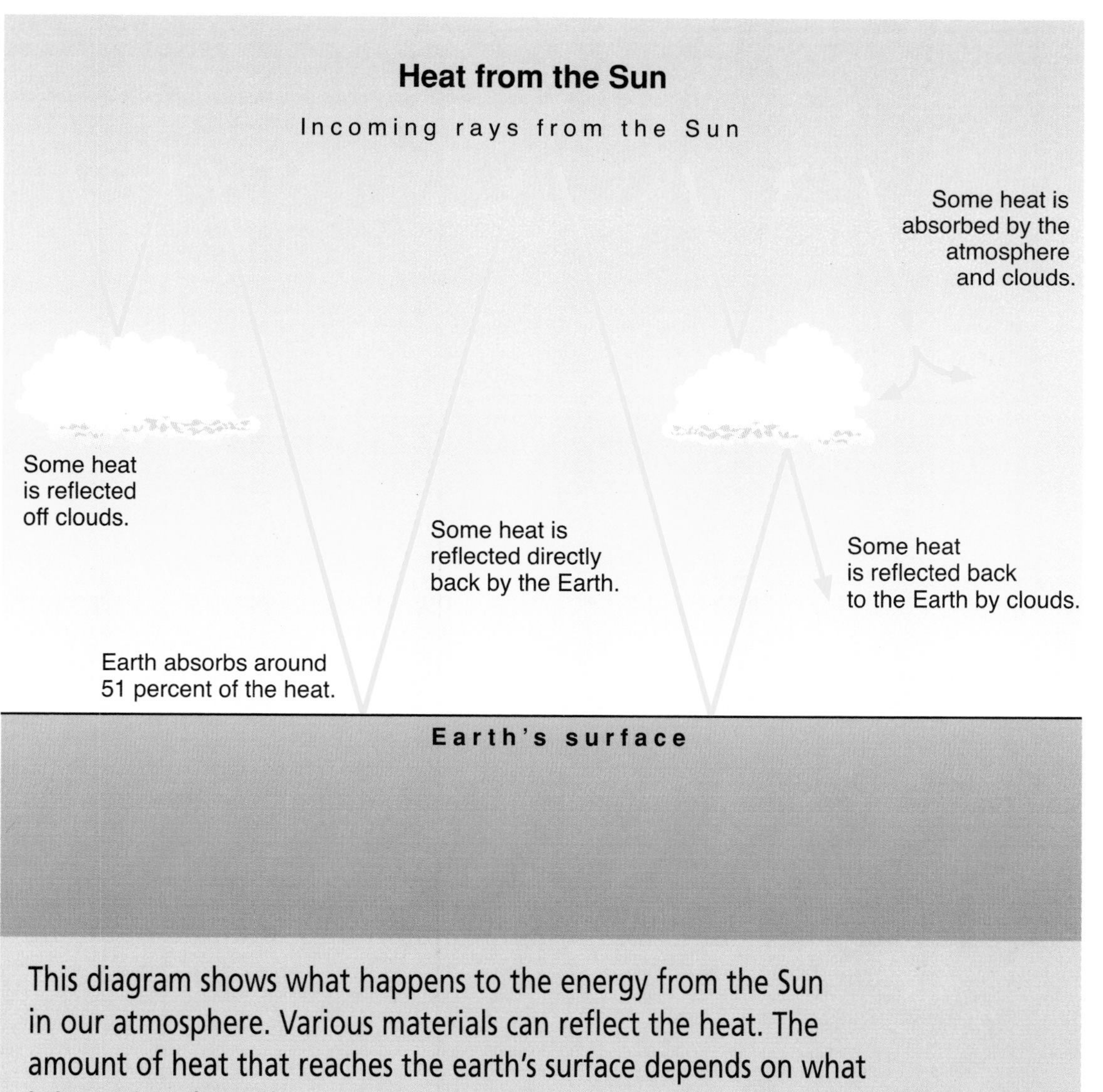

This diagram shows what happens to the energy from the Sun in our atmosphere. Various materials can reflect the heat. The amount of heat that reaches the earth's surface depends on what it is passing through.

Day and Night

As the earth rotates, continents travel into sunlight and then into shadow. This is what causes us to experience day and night. There are large temperature differences between night (which is usually cooler) and day (which is usually warmer). Also, land masses and oceans heat up and cool down at different rates. During the day, land masses warm up more quickly than oceans do. At night, oceans cool down quite slowly compared to land. On a warm summer's day, this temperature difference produces a breeze that moves from the sea to the land. At night, after a sunny day in the summer, the breeze moves from the land to the sea.

Clouds have a large effect on the temperature differences between day and night. During the day, they stop sunlight from reaching the earth, making it cooler. At night, they act like a blanket and keep in the warmth of the earth. Clouds stop this warmth from escaping into space, which prevents the temperature from falling sharply.

Day and night can be clearly seen in this image. It is midday on the far right of the picture and midnight on the far left. The part of the earth nearest to us is moving toward the right. Is it sunrise or sunset on the line dividing day from night?

Day, night, and clouds

On a warm sunny day, heat coming off the earth can force air upwards in **thermals.** If this air contains enough moisture, it will form fluffy **cumulus** clouds when it reaches a cold enough layer of air. These clouds appear and disappear on a warm sunny day. At night, there is no sunlight to warm the land and make thermals, and you will rarely see cumulus clouds. Cumulus-type clouds do not form in Antarctica, as the very cold temperature on the ground does not produce thermals.

Dew and frost

When there is no sunshine or clouds, it can be cold enough for frost and fog. After the Sun has gone down, it gets cooler and the temperature of the ground falls. As the temperature of the air falls, drops of water called dew form on the ground. Dew is formed when the **water vapor** in the air condenses as it touches cool surfaces. If it is very cold, the drops of water will turn straight into **ice crystals.** When this happens we have a **hoar frost.**

These drops of water were not caused by rain. Water vapor in the air has turned back into water as the temperature has dropped. It is called dew.

UV Radiation

We can all see the light that comes from the Sun, but the Sun's rays also contain ultraviolet (UV) **radiation.** This UV radiation—which we cannot see—contains a lot of energy. Some of it is harmful to living things like humans. Almost all of the most harmful UV radiation is absorbed on its journey through the **atmosphere.** The rest reaches the earth.

UV radiation causes skin to react and produce a tan to protect itself. A tan is a temporary darkening of the skin. The darker your skin, the fewer UV rays are absorbed by your body. Some people tan more easily than others. Those who do not tan easily may have their skin burned more quickly by UV radiation. Even those who do tan may receive a harmful dose of UV rays while tanning. UV radiation causes skin cancer, and can make skin wrinkle at an earlier age. While it may look attractive to have a tan, it is not worth risking your health for.

The UV Index

The UV Index shows what form of protection you need against the Sun. The readings are given on a scale of 1 to 10+. Weather forecasts on television and in newspapers include maps showing these readings. Use the chart to find out whether you are likely to damage your skin or not. "Low" means that there is little risk of burning or damaging your skin. "Very high" means that it is almost certain that you will burn, so you should stay out of the sun.

Exposure levels and their risks to different skin types					
UV Index number	**Exposure level**	**Light skin that burns**	**Light skin that tans**	**Medium skin**	**Dark skin**
1–2	Minimal	Low	Low	Low	Low
2–4	Low	Medium	Low	Low	Low
5	Medium	High	Medium	Low	Low
6	Medium	Very high	Medium	Medium	Low
7	High	Very high	High	Medium	Medium
8	High	Very high	High	Medium	Medium
9	High	Very high	High	Medium	Medium
10+	Very high	Very high	High	High	Medium

Be safe!

Avoid the effects of UV radiation:

- Don't spend too much time in the sun between 10:00 A.M. and 4:00 P.M.
- Work or play in the shade whenever you can.
- Always wear sunscreen of at least SPF (Sun Protection Factor) 15—even in the shade.
- Wear a hat with a wide brim.
- Wear tightly woven, full-length clothing.
- Wear UV protective sunglasses.

Remember that even when you are in the shade, sidewalks, grass, water, and sand can reflect the Sun's UV radiation onto you. How is this boy protecting himself from the Sun's strong rays?

What Are Clouds?

People are usually interested in which clouds bring rain. However, there are many types of clouds, and not all of them bring rain. Knowing about clouds makes it easier to forecast the weather.

A cloud is formed when moist, or damp, air cools. Weather happens in a particular part of the **atmosphere.** The higher you go in this part of the atmosphere, the cooler it gets. In air with little moisture in it, there is about a 1°C drop in temperature for every 328 feet (100 meters) you go up. As the air cools, the **water vapor** in the air condenses around tiny particles of salt or **pollutants,** forming water droplets. Along with billions of others, these droplets make up a cloud.

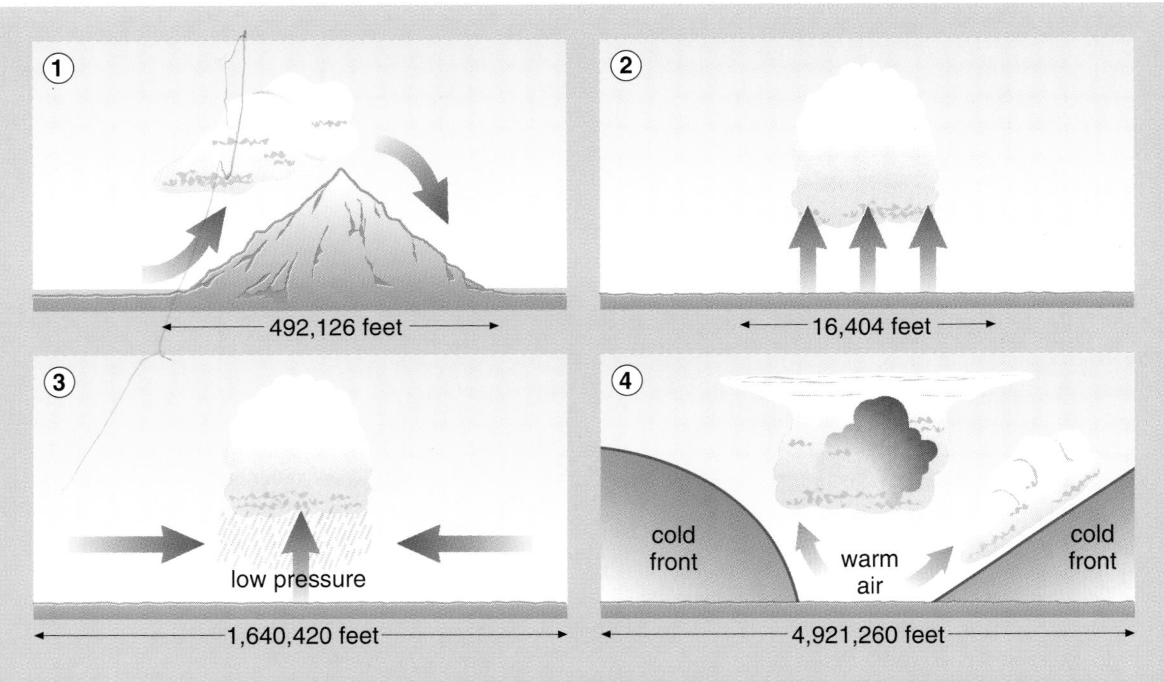

There are four main ways in which clouds are caused. All of them involve the water vapor being forced higher in the sky where it is cooler.

1. Relief topography, or mountain cloud: Clouds and rain can be produced by moist air being forced up by obstacles such as hills or mountains.
2. **Convectional** clouds: Heat rising from the ground forces moist air up. These events are called **thermals** and are a good source of clouds. They are called convectional clouds because rising heat is known as convection.
3. **Convergence** of air: Clouds are formed when air meets from opposite directions and is forced upwards. This is called convergence.
4. Lifting air along **weather fronts:** Weather fronts, which occur when different temperature **air masses** meet, force air upwards. Clouds are produced and they often bring rain.

If there are so many ways in which clouds are produced, how do we ever see clear blue skies? When the sky is blue, especially in the summer, there may be high **air pressure.** High air pressure means that the air above is sinking down. This means that air cannot rise and cool, so clouds will not form.

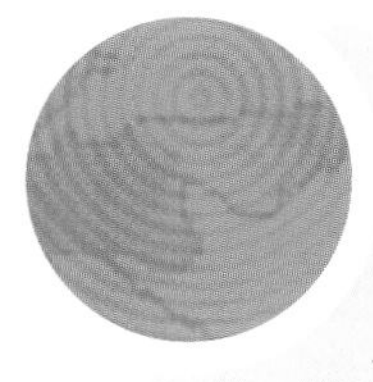

Try this yourself!

You can make your own cloud in a large plastic bottle. (Make sure you ask an adult to help you with this project.)

- Put some water into your container. Shake the bottle to mix the air and water.
- Get an adult to light a match and put it into the container. The water will extinguish the flame.
- Tighten the lid. Squeeze and release the bottle. What do you see?

Shaking the bottle mixes air and water, putting water vapor into the air. The smoke provides tiny particles for the cloud to form around.

Cloud Cover

When recording the weather, there is a big difference between a sky with a few small clouds and an **overcast** sky. There are several ways of recording the total amount of cloud cover. One popular system is based on mentally dividing the whole visible sky into eighths or "oktas" (divide the sky in half, in half again, and then in half once more). In the United States, the system is based on tenths of the sky.

The table below shows how cloud cover can be calculated. The terms in the right-hand column provide additional information that non-experts can use. On weather forecasts, these are usually illustrated with sun and cloud symbols which show how much sun and cloud there is.

Cloud Cover			
Amount of sky covered	**Oktas (eighths)**	**Tenths**	**Description**
Clear sky	0	0	Sunny
One quarter of sky	2	2 to 3	More sunny than cloudy
One half of sky	4	5	Partly sunny
Three quarters of sky	6	7 to 8	More cloudy than sunny
Full sky	8	10	Overcast

It is surprising how much of the sky is obscured if you are standing near a building. This could give an inaccurate idea of how much cloud cover there is. Walk out into an open space and look all around, concentrating on how much cloud cover is in the sky.

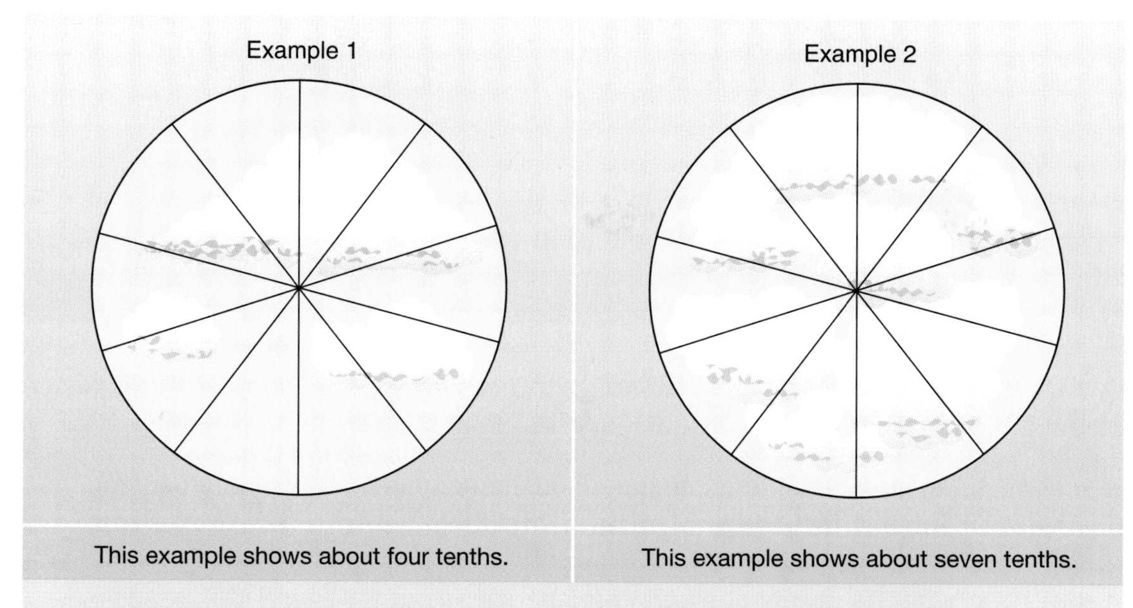

This example shows about four tenths.

This example shows about seven tenths.

These diagrams give examples of the sky as you might see it. The cloud can be scattered all over the sky. The lines shown here show you how to divide the sky up into the correct number of portions. To work out the cloud cover you need to decide how many of them would be full if all the clouds were put together in one place.

Recording cloud cover

Cloud cover is measured as close to 9:00 A.M. as possible. If you measure cloud cover at the same time each day, you will have a better idea of what the cloud cover patterns are over a long period of time. When recording cloud cover, you must stand where there is a clear view of the sky. Stand away from buildings and other obstacles. If the sky is hidden by fog, record that there is **obscured** sky.

If the clouds are spread out in the sky, try to imagine how much of the sky they would cover if they were placed all together. This takes a little practice. Remember not to look directly at the Sun. You can take two recordings each day. In the winter, the second recording should be taken at night. In this case, you could figure out the amount of cloud cover by looking at the parts of the sky where you can see stars.

In some places, the light from street lights reflected on the clouds shows us the areas that are covered by clouds. Professional **meteorologists** shine spotlights or lasers on the clouds to figure out how high they are.

Cloud Classification

In science, it is important to name things. In 1803, Luke Howard, an English scientist, suggested names for the different types of clouds. His system had four basic cloud forms, which used Latin words to describe what he saw: **stratus** (layers), **cumulus** (heaps), **cirrus** (wisps of hair), and **nimbus** (violent rain). These words could be combined into a single name. For example, stratocumulus is a layered and lumpy cloud. In 1887, two scientists called Abercromby and Hildbrandsson expanded Howard's system. This is the system **meteorologists** still use today.

There are ten main types of clouds. Each type is classed into one of the three main height levels that clouds are found at in the sky. These are high, middle, and low clouds. Cumulonimbus clouds can be seen at more than one level.

Luke Howard created his system for naming clouds after studying them for a long time. His system makes understanding cloud names much easier. If a separate name had to be learned for each of the ten clouds, it would be much harder.

Cloud types

The ten main cloud types are shown, with their abbreviations, in the chart below.

<table>
<tr><th colspan="5">Main Cloud Types</th></tr>
<tr><th>Height
(in feet)</th><th>Stratus clouds
(layers)</th><th colspan="2">Cumulus clouds
(heaps)</th><th>Other clouds</th></tr>
<tr><td>High level
16,400–42,650
(cirrus)</td><td>Cirrostratus (Cs)</td><td>Cirrocumulus (Cc)</td><td rowspan="5">Cumulonimbus (Cb)</td><td>Cirrus (Ci)</td></tr>
<tr><td rowspan="2">Middle level
6,560–22,970
(alto)</td><td>Altostratus (As)</td><td rowspan="2">Altocumulus (Ac)</td><td rowspan="2"></td></tr>
<tr><td>Nimbostratus (Ns)</td></tr>
<tr><td rowspan="2">Low level
0–6,560
(stratus)</td><td>Stratus (St)</td><td>Cumulus (Cu)</td><td rowspan="2"></td></tr>
<tr><td colspan="2">Stratocumulus (Sc)</td></tr>
</table>

The first part of the names of these clouds may help you figure out how high they are in the sky. The rest of the word may help you figure out the way the cloud looks. The abbreviations that meteorologists use on their data recording sheets are in parentheses. It would be hard for them to fit the full names of the different cloud types into the tiny boxes on their sheets! You can find out more about these clouds on the following pages.

How high are clouds?

It is difficult to work out the height and size of clouds. This is especially so with cirrocumulus and altocumulus clouds. Experienced weather watchers use the fingers of their outstretched arm as a guide. If you can cover individual clouds with just one finger, you are looking at cirrocumulus clouds. These are small clouds which look even smaller because they are so high up. If it takes three fingers to cover a cloud you are looking at altocumulus clouds. These are larger and can be found at the middle-level height.

At the **polar** regions, each cloud type is found much lower in the sky than shown in the chart. In the **tropical** regions, each type is found much higher.

Cumulus Clouds

When young children are asked to draw a picture of clouds, they often draw a lumpy shape. This is what cumuliform, or **cumulus,** clouds look like. They are made up of small heaps of cloud, with clear edges and flat bottoms, which show up well against the blue sky. Cumulus clouds are not always white. They can also be many shades of grey. Cumulus clouds are often formed on hot summer days, when heat rises and reaches the cold air higher up. These clouds disappear as they **evaporate** again.

When they are low in the sky, cumulus clouds have three basic shapes. Cumulus humilis is flat-bottomed and wider than it is tall. It is a fair-weather cloud. However, it can grow into cumulus mediocris. This cumulus cloud is as tall as it is wide. This might grow into cumulus congestus, which is taller than it is wide. Cumulus congestus can produce moderate to heavy rain. They can also be signs that cold weather is approaching. Beware: they may be followed by cumulonimbus storm clouds!

Fires can cause clouds of **condensation** to appear in the sky. When a fire is burning, smoke mixes with clouds to make special clouds called pyrocumulus clouds. *Pyro* comes from the Greek word for fire.

Sometimes, cumulus clouds are very high up in the sky and grouped together into waves of clouds, called cirrocumulus. Each individual cloud is quite small. They are so high in the cold sky that they are not made of **water vapor,** but of **ice crystals.**

Bad weather cumulus clouds

You would not want to spend too long outside under cumulus clouds that have joined together into huge coverings of clouds called stratocumulus clouds. These are dull blankets of cloud that form when the tops of the clouds rise and spread out sideways. They will bring rain or snow showers.

Not all cumuliform clouds are harmless. Some can develop into rain-bearing cumulonimbus clouds. These clouds bring violent hail, thunder, and lightning, as we shall see later in this book (page 26).

The altocumulus, or mackerel sky, is made of layered clouds with waves that are like the pattern on the mackerel fish. These clouds sometimes show that a change in the weather is approaching.

Stratus Clouds

Stratiform, or **stratus,** clouds are made up of layers of unbroken clouds that have a flat base. These often appear to be white in the sky. Nimbostratus, altostratus, and cirrostratus clouds appear in the lower, middle, and upper parts of the sky. Stratus clouds are found at all levels of the sky. When a stratiform cloud appears in the lowest level of the sky, it brings drizzle or, if it is cold enough, snow.

Stratus clouds can be formed from moist air that has not risen very far into the sky. Sometimes, buildings on hills can be covered by stratus clouds! This shows how low to the ground they can be. Stratus clouds can be so thin that they do not block out the Sun. Because of this, care should be taken not to look directly at the sun through this cloud. Sometimes, this low cloud becomes lumpy and patchy all over. Then it is called stratocumulus. This is a layer of **cumulus** clouds.

Stratus is both a cloud itself and a type of cloud. This grey sky is covered by a layer of stratus clouds, which are situated in the lowest level of the **atmosphere.** Stratus clouds can bring drizzle.

Other types of stratus cloud

Cirrostratus is a very high cloud. It is cold and made out of **ice crystals.** It is very pale, smooth, and thin. When cirrostratus only covers part of the sky, the other part of the sky often has other types of **cirrus** clouds.

Altostratus is found in the middle level of the sky. It is thicker than cirrostratus and is made of water droplets. The cloud is greyish or bluish in color. It can be thin enough for the Sun to shine weakly through it, but it covers the Sun enough to stop it from casting any shadows.

If the cloud is so thick and dark that it blocks the Sun, and if rain or snow is falling, then the cloud is a nimbostratus.

Cirrostratus clouds are very thin. The beautiful circle made by the moonlight shining through shows that they contain a fine layer of ice crystals. These clouds are so thin that this **halo** around the moon is sometimes the only clue that the clouds are there. Altostratus clouds are far easier to see. They are lower in the sky and are made up of water droplets instead of ice crystals. They do not produce halos.

Cirrus Clouds

Cirriform, or **cirrus,** clouds can be among the most beautiful of clouds to look at as they flow and form in the sky. On a nice day, they are strong indicators that a change in the weather is coming.

Cirrus clouds are found high up in the sky, where the temperature is extremely cold and the winds are very fast. They are made up of water, in the form of millions of **ice crystals** that are stretched out across the sky. The direction of the streaks of the clouds shows which way the wind is blowing at this level of the sky. The word cirrus comes from Latin. It means "'wisp of hair," which is what some people thought this type of cloud looked like.

Cirrus uncinus is a beautiful type of cirriform cloud. These clouds were thought to look like the tails of female horses. They were given the name "Mares' Tails" because of this.

Types of cirrus cloud

One type of cirrus cloud is cirrus uncinus. At the end of these streaky clouds are what look like upturned hooks. These are formed when ice crystals high in the air begin to fall, but are then blown into long streaks by strong winds below them. The "hooks" show the place where the crystals began to drop. The streaks show the direction in which the wind was traveling.

Cirrostratus is a high, very even, thin-layered cloud. It often covers much of the sky. There are two types of cirrostratus clouds. Cirrostratus fibratus is created by continuous strong winds. It is long, and has thin lines that spread out across the sky. Cirrostratus nebulosis is made by gentle rising air. This cloud is hard to see, but when the Sun shines through it at the right angle you can see a **halo** around the Sun.

Sometimes you can spot airplanes easily in the clear blue sky. This is because their exhaust fumes produce long, thin, white cirrus clouds called contrails. These often disappear quickly, but they can last up to 30 minutes when there is a lot of moisture high in the sky.

These are man-made clouds. They are formed when **water vapor** from aircraft engines condenses. Sometimes planes make holes in clouds. The holes are called distrails.

Giant Clouds

The real giant of the cloud kingdom is cumulonimbus **incus.** It is the biggest cloud of all. The storm that it brings can release as much energy as an atomic bomb. This cloud grows in stages. Sometimes cold, heavy air forces warm air to rise up quickly. This causes a rapid and violent change in the weather. This is when **cumulus** clouds are formed. These may then develop into giant clouds.

Clouds grow and develop. These clouds will only develop if warm air continues to rise:
1. cumulus humilis **2.** cumulus mediocris
3. cumulus congestus **4.** cumulonimbus calvus
5. cumulonimbus with **pileus**
6. cumulonimbus incus.

Giant clouds and the weather they bring

Cumulus humilis (1) and cumulus mediocris (2) wouldn't spoil a picnic! However, cumulus congestus clouds (3) develop when there are strong **updrafts** of warm air with cold air above, which cause **turbulence.** Cumulonimbus calvus (4) are big, white clouds up to 29,528 feet (9,000 meters) high. These clouds bring moderate to heavy showers and strong winds.

Cumulonimbus with pileus (5) are giant clouds 19,685 to 29,528 feet (6,000 to 9,000 meters) high. If updrafts of air grow stronger, they can blow the top off a cumulonimbus calvus cloud, producing a cap, or smooth flat top. The cumulonimbus calvus cloud then grows and catches up with its "cap." This cloud means the weather will turn nasty.

Cumulonimbus incus clouds (6) bring rain, strong winds, and even tornadoes. At 59,055 feet (18,000 meters) tall, they are so big they would make Mount Everest look small! Hail is also produced in this cloud. Hail is made of ice particles that travel up and down inside the cloud. They melt and refreeze until they become so heavy that they fall to the ground.

Cumulonimbus incus clouds also produce thunder and lightning. Negative and positive electrical energy travels back and forth between the cloud and the ground. This is lightning. Lightning traveling through the air makes the bang of thunder. We see the lightning before we hear the thunder. This is because the speed of light is much quicker than the speed of sound.

Be safe in a thunderstorm!

- Stay indoors, or in a car (but don't touch its sides).
- Do not use the Internet or the telephone.
- If you are outdoors don't seek shelter under a tree, as lightning strikes tall objects.
- Because lightning strikes tall things, get lower to the ground than the object that is nearest to you.

Fog and Visibility

Fog and mist are also types of clouds. When the air is full of **water vapor,** some of it will turn into water droplets. These form around tiny bits of dirt in the air. Fog often occurs in the morning or evening, until the warmth of the sun **evaporates** it. The water droplets in fog are the same size as those found in clouds. The water droplets in mist are much smaller. This explains why mist is not as thick as fog. If you cannot see as far as 0.6 miles (1 kilometer) away you are in fog. If you can see farther than that distance, you are in mist.

Four kinds of fog-type clouds

Radiation fog is caused when the air above rivers, lakes, and valley bottoms cools down to the point (called the dew point) where the water in the air condenses and becomes fog. This type of fog often happens on cool, clear nights when the heat has left the ground. If the weather stays cool this fog can last for days, but in sunny weather it is cleared by the warmth of the Sun's rays.

The cooling of air near the surface of the water produces this radiation fog. If it gets warmer, it will soon disappear.

Advection fog (sometimes called sea fog) is created when warm, moist air travels over a cold land surface. It also occurs when the same type of air passes over a cold current at sea. This fog does not rise above about 1,640 feet (500 meters) from the ground, as the temperature starts to get warmer above this. This is why the top of the Golden Gate Bridge in San Francisco often peeks out over the top of the fog.

Steam fog, or "arctic sea smoke," is the result of cold air hanging over warm water. For this fog to form, there needs to be a temperature difference of at least 50°F (10°C) between the air and the water. This fog is often seen in the Arctic, as very cold air blows over warmer water until the water looks like it is steaming. It can also be seen on roads that are warmed by the Sun after a sudden shower of rain.

When there is a lot of pollution in the air, a nasty type of fog called smog is produced. The only way to stop smog is to control chemical **emissions** from vehicles and factories.

Four Types of Fog

	When	Where	Air/Surface Conditions	Wind
Radiation Fog	Nighttime	Rivers, valleys, low places, not over sea	Cooling ground, moist air	Light breeze
Advection Fog (sea fog)	Daytime	At sea	Warm moist air over cold surface	Gentle wind
Steam Fog	When there is at least 50°F difference between water and air	The Arctic; on sun-warmed roads	Cold air over warm water	Still air
Smog	When smoke and pollutants are present in the air	Industrial areas	Layer of warm air over layer of cold air	Still air

Glossary

air mass body of air in which all the air is of approximately the same temperature and humidity

air pressure pressure at the surface of the earth caused by the weight of the air in the atmosphere

atmosphere the gases that surround our planet. They are kept in place by Earth's gravity.

big bang theory stating that the universe began in a huge explosion

cirrus highest form of clouds, made up of ice crystals in thin, feather-like shapes

climate pattern of weather in an area over a long period of time

condensation water that has changed from a gas into a liquid, or the process of changing from gas into liquid. When a gas turns into a liquid, it condenses.

convectional vertical movement, especially upwards, of warm air

convergence meeting at one point

cumulus type of cloud consisting of rounded heaps with a flat bottom

data group of facts that can be investigated to get information

ellipse oval shape

emission substance blown out into the air, usually by a factory or an automobile

equator imaginary horizontal line around the center of the earth, at equal distance from the north and south poles

evaporate when water changes from a liquid into water vapor

fossil fuel natural fuel such as coal, oil, or natural gas

greenhouse effect effect of gases that act as a kind of insulator, stopping the earth's warmth from escaping. The term is also used to describe the increase in gases that has caused too much heat to be kept in the atmosphere, warming the earth.

halo circle of light around a light source

hoar frost ice that forms on surfaces near the ground when the temperature falls below zero

ice crystal small piece of frozen water

incus Latin word for anvil; iron block on which metals are hammered out. It is used to describe the flattened, anvil-like shape at the top of some cumulonimbus clouds.

meteorologist person who studies the weather by gathering and analyzing data

nimbus large grey cloud that brings rain

obscure to block

overcast when the sky is completely covered by clouds

pileus Latin for soft felt cap. A pileus cloud has the shape of a flat hat and occurs at the top of, or above, some cumulonimbus clouds.

polar having to do with the north or south pole

pole imaginary point found at the most northern and southern places of the earth. Both have very cold climates.

pollutant something that pollutes

radiation type of energy

sphere round, ball-like shape

stratus type of cloud—made up of layers—that has a flat base

thermal current of rising heat

tropical having to do with the region around the equator

turbulence when air moves around, causing winds and drafts

updraft rising current of air

water vapor water in the form of gas

weather front location where air masses of different temperatures and humidity meet. This is often the place where the most significant weather happens.

More Books to Read

Dunn, Andrew. *Fog, Mist and Smog.* N.Y.: Raintree Steck-Vaughn, 1998.

Gold, Susan Dudley. *Blame It on El Niño.* N.Y.: Raintree Steck-Vaughn, 1999.

Harper, Suzanne. *Clouds: From Mare's Tails to Thunderheads.* Danbury, Conn.: Franklin Watts, 1997.

Oxlade, Chris. *The Weather.* N.Y.: Raintree Steck-Vaughn, 1999.

Trueit, Trudi Strain. *Clouds.* Danbury, Conn.: Franklin Watts, 2002.

Index